COACHBUILDING
THE HAND-CRAFTED CAR BODY

Jonathan Wood

SHIRE PUBLICATIONS

Published by Shire Publications Ltd,
PO Box 883, Oxford, OX1 9PL, UK
PO Box 3985, New York, NY 10185-3985, USA
Email: shire@shirebooks.co.uk   www.shirebooks.co.uk

First published 2008.
Transferred to digital print on demand 2014.

A CIP catalogue record for this book is available from the
British Library.

Shire Library no. 476   •   ISBN-13: 978 0 7478 0688 2

Jonathan Wood has asserted his right under the Copyright,
Designs and Patents Act, 1988, to be identified as the
author of this book.

Designed by Ken Vail Graphic Design, Cambridge, UK
Typeset in Perpetua and Gill Sans.
Printed and bound in Great Britain.

COVER IMAGE
The bodyshop at Rippon, Yorkshire (see page 30).

TITLE PAGE IMAGE
A 1909 20 hp A-type Vauxhall enhanced with a magnificent
landaulet body, parked outside the premises of its
coachbuilders, W. and F. Thorn of 19–21 Great Portland
Street, a centre of the London motor trade. The customer
was the Duke of Northumberland, Vauxhall's first titled
client, and the body finished in his colours of blue, white
and yellow. The car company's own offices in the capital
were nearby at numbers 180–2.

CONTENTS PAGE IMAGE
An H. J. Mulliner drawing of a sedanca coupé on a Bentley
chassis, one of 102 bodies it built on the 4.25 litre chassis.
Its stylist and works director was Stanley Juniper Watts.

ACKNOWLEDGEMENTS
I have had much help in the preparation of this demanding
title, and first and foremost my thanks must go to the
Rolls-Royce Enthusiasts' Club for staging an excellent
seminar on coachbuilding. The papers by Roy Brooks on
Joseph Cockshoot, Tom Clarke (Gurney Nutting) and Will
Morrison (Mulliner) greatly added to my knowledge of the
subject. I also consulted Roy Brooks's MSc thesis on
Cockshoot, held in the library of the Sir Henry Royce
Memorial Foundation, where Philip Hall, its chief
executive, offered every assistance.

Thanks are likewise due to the staff of the National
Motor Museum's library and I was made equally welcome
at the library of the Vintage Sports-Car Club.

I should like to record particular thanks to Tom Clarke,
who read the manuscript and made helpful suggestions,
although I alone am responsible for its contents. I also
appreciate the help I received from Robin Barraclough,
Harvey Cooke (Post Vintage Humber Car Club), John
Dyson and Will Morrison of MotorHistorica.

Illustrations were kindly supplied by: Clive Baker
(Hillman Register), 26 (lower); John Bath (Triumph
Razoredge Owners' Club), 51; Bill Bertelli, 24; Barry
Blight 32 (uppers), 33 (middle), 38; Lionel Burrell (The
Automobile), 26 (upper); Tom Clarke, 3, 37 (upper), 42
(both); Bryan Goodman, 10, 11 (upper and lower), 21
(upper), 27, 37 (lower), 39, 44 (lower); Malcolm Jeal, 8
(lower), 9 (lower); The Morgan Motor Company, 52
(lower); Jacque and Gwyn Morris (Riley RM Club), 50;
Will Morrison 22, 24 (upper), 40 (upper); Ann Pilgrim
(Rapier Register), 43; Nic Portway 33 (lower); Peter
Seymour, 28 (upper and lower), 29. The remainder are
from the author's collection.

Shire Publications is supporting the Woodland Trust, the UK's leading woodland conservation charity, by funding the dedication of trees.

CONTENTS

CONTINENTAL CURTAIN-RAISERS, 1886–1914

COACHBUILDING is not only as old as the motor car, which dates from 1886; it predates it by at least three hundred years. Today the term relates to the traditional construction of car bodywork in which wood and metal are triumphantly united to create a shape that is wholly reliant on the skill of the craftsman producing it. This craft originated in the carriage trade that flourished throughout Britain and Europe in the eighteenth and nineteenth centuries.

The horse-drawn carriage was essentially a Continental concept and many of the manufacturing processes, styles and terms originated south of the English Channel. Indeed, the word *coach* is a derivative of the Hungarian *kocsi*. As coachbuilder William Bridges Adams observed in 1837, 'the English carriage builder takes a French or German carriage to improve upon, because it saves his time and trouble...' His twentieth-century successors, makers of automobile coachwork, would perpetuate this approach.

Much of the nomenclature that applied to carriages was used to describe motor-car coachwork, even if the resemblance between the two was sometimes elusive. But the use of an established name provided continuity and was thus a strong marketing consideration. Among the best-known terms were: brougham, cabriolet, coupé, dickey, dog-cart, landaulet, phaeton, sociable, victoria and wagonette.

When motor cars began to appear in Britain in the 1890s they were described as 'horseless carriages', a translation of the French expression '*voitures sans chevaux*', indicative of a country that had already embraced the automobile.

However, the motor car was a German invention propagated by the French, and many of the stylistic initiatives found on British cars were first seen amongst the potted palms of the Paris Salon, first staged in 1898.

Opposite:
It was a simple step to add a roof to an open body, although the chauffeur lacks weather protection on this Peugeot, c.1905. The rear occupants have the advantage of doors, but there is none at the front.

5

Daimler's Mercedes marque of 1901 popularised the honeycomb radiator. Note that the simple bodywork of this victor of the French Grand Prix incorporates a scuttle in the manner of passenger cars. It is one of a series of art nouveau tile paintings which commemorated Michelin Tyres' competitive successes and enhanced its one-time London premises, opened in 1910.

But during the motor car's journey across the English Channel Gallic exuberance was replaced by an anglicised restraint.

The power units that propelled the automobiles of Carl Benz and Gottlieb Daimler of 1886 were single-cylinder engines located amidships, behind the driver and passenger. In the first instance these automobiles possessed no bodywork in the accepted sense although they were 'cart

The bodywork on this 1898 Benz is minimal. Note that the front wheels, which pivoted centrally, axle and all, are smaller than the rear ones, an inheritance from the horse-drawn carriage.

sprung' on full elliptic springs, an inheritance from the horse-drawn carriage.

By the time Benz had introduced his victoria in 1893 it had acquired a hood and rudimentary wings and was a four-wheeler, his previous vehicles having had only three wheels. It also followed carriage practice in that its front wheels were smaller than the rear ones. This was an inheritance from the days when the centre-pivoting front axle had to pass under the coachman's box as the carriage turned a corner.

Motor cars soon adopted another legacy of the horse-drawn era, albeit a little-used one, namely Ackermann steering, in which the axle remained stationary, and each wheel turned on its own stub axle and was connected to the other wheel by a track rod. In consequence carriage-style wheels of differing diameters were no longer required and after about 1900 same-sized wheels became the norm.

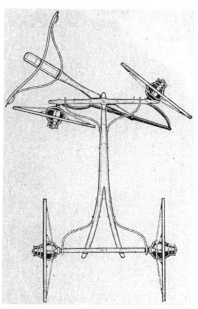

The accommodation problem was addressed, in the short term at least, by the introduction of a secondary bench behind the driver, the resulting body being named the double or Siamese phaeton. But because cars had short wheelbases and their wheels were large, the luckless passengers were perched dangerously high.

More satisfactory was the rear-entrance tonneau, arguably the first body style to be designed specifically for the motor car. *Tonneau* means a tub or cask in French and the style arrived in about 1897. It was now possible to accommodate a maximum of three additional passengers, usually facing forward, behind the driver. One passenger sat in each corner of a wooden tub with the third perched on a seat attached to the door itself. This type of tonneau survived until 1904–5, its limitations being that the occupants might have to alight into a muddy road and it was difficult to provide protection from the weather.

In the meantime the motor car's mechanicals had undergone a significant realignment. In 1891 Emile Levassor of the French Panhard et Levassor concern moved the engine, in this instance a vertically positioned Daimler V-twin, to the front of the vehicle. With it came a primitive cooling element, later to acquire a distinctive shell, and a rudimentary bonnet.

Drive thereafter passed, via a clutch, to a gearbox, with final drive to the rear axle being by chains. What had emerged was a motor car, as opposed to a horseless carriage. Here was a concept that could be subject to almost unlimited refinement, and *système Panhard* was destined to dominate global automobile design for the next seventy or so years.

The carriage trade bequeathed Ackermann steering to the automobile, so permitting front and rear wheels of the same diameter to be used. The inventor, in 1816, was a German coachbuilder, George Lenkensperger of Munich, but the device takes its name from the publisher Rudolph Ackermann, who in 1818 acquired the English rights.

This 10 hp Panhard, c.1902, well illustrates the problems experienced by early coachbuilders when the chain drive prevented access to the rear seats. The answer was the tonneau body, with passengers entering via a central door located at the rear of the vehicle.

This configuration initially applied to the finest and most influential manifestation of the line, the Roi des Belges body, which appeared in the spring of 1901. The style originated at a meeting between King Leopold II of Belgium, his mistress, Cleo de Merode, and Ferdinand Charles, of coachbuilders J. J. Rothschild & Sons Parisian, at her apartment in Avenue Louise, Paris. During the discussion the lady put together two of her tub armchairs. They were upholstered, according to the pioneer motorist and coachwork designer Montague Graham-White, 'in rich morocco leather,

The first Roi des Belges body of 1901 for the King of Belgium on a 20 hp Panhard et Levassor chassis in its original colours. Note how only the front seats resembled armchairs (see text); the monarch's, to the left of the driver, was of extra width.

thickly padded, with heavy "roll over" trimming, pleated and buttoned'. She suggested that they form the basis of the front seats. 'The idea pleased the king and he suggested that Mr Charles should make out a rough sketch on the spot, particularly in following the tulip-shaped lines of the chairs in front of him.'

Mounted on a 20 hp Panhard chassis, the *Roi des Belges Tonneau de Grande Luxe,* to accord the design its full title, was replete with the Art Nouveau curves so appropriate to Paris. Then came another significant structural innovation, because the body was panelled not in wood, so beloved by carriage makers, but in aluminium, a metal then newly but widely available. This was because, as a correspondent observed, 'one cannot persuade a wooden plank to assume a concavo-convex form'. The result was the composite body that was destined to become the norm, although steel was initially to be a more popular alternative. Visually exhilarating, Charles's lines were widely imitated and enjoyed great popularity until about 1910.

Initially rear access was a problem, with passengers invariably gaining entry via a swivelling nearside front seat. The side entrance began to emerge in about 1905. Access to the rear seats was, at last, by doors, made possible by the adoption of shaft drive, so removing the intrusive chains, increasing wheelbases, reducing the diameter of the wheels and moving the rear axle closer to the back of the vehicle. By this time the four- and five-seater tourer was predominant, consolidating the popularity of the phaeton.

It was but a short step to add a roof to the phaeton and thus create the limousine, the first examples appearing during the 1902–3 seasons. The name, unlike so many of the day, was peculiar to the motor car and was derived from a coarse woollen cloth applied by French carriers to protect the contents of their wagons from the sun. The cloth was named after the

The Roi des Belges body was widely copied, the lines having been extended to the rear seats. Their tulip-like contours are very obvious on this 1905 Regina, the name under which the German Dixi was sold in France. The occasion was the 1905 Loiret Cup.

9

There is no windscreen on this Chenard-Walcker 16/20 of 1906. The body in the Roi des Belges style was described as a victoria phaeton. Note the absence of a scuttle.

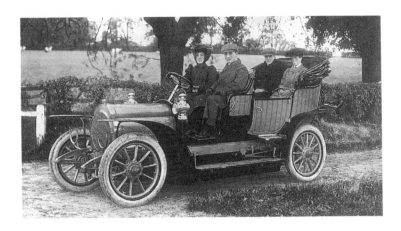

Limousin region of south-west France, where it was made, part of the country well-known for its weaving.

The landaulet, a derivative influenced by the horse-drawn landau, offered the advantages of both the limousine and the phaeton. If the weather permitted, the rear portion of the body could be folded down so that the back-seat passengers could enjoy the fresh air. Lofty, to permit gentlemen in their 'toppers' and ladies in voluminous hats to enter without removing their headgear, the roofs of these bodies could reach a height of 7 feet.

Such coachwork underpinned the rigid social structure of the day by incorporating a division between the chauffeur and the car's occupants. The driver was only a little better off than his coachman predecessor, who would have been exposed to the elements, but initially he would not have benefited from the protection of a windscreen, side windows or even doors.

This contrasted with the often sumptuous, coach-like interior for the master and his lady, which might be comfortably upholstered and enhanced with the finest veneers. Windows were operated by leather straps, in the manner of old-fashioned railway carriages or indeed the coaches from which they had derived.

It was customary for owners to use their limousines and landaulets about town, for shopping, or attending the theatre and restaurants. However, phaetons were preferred for annual pilgrimages to the Riviera during the winter months, when owners sometimes took the wheel. Such an activity was less acceptable in towns, a line of demarcation reflected by an advertising slogan of the day for a particularly imposing French marque. It cautioned: 'One doesn't drive a Delaunay-Belleville – it simply isn't done.'

Massive mahogany-framed plate-glass windscreens had begun to appear on open and closed cars in about 1902. They were mounted directly above the 'dash-board', this spelling being the original rendering. It was another

The carriage origins of the Labourdette double limousine bodywork on this Lorraine-Dietrich remain unsullied, even by 1910. The chauffeur, unusually, was protected from the elements.

term inherited from the carriage trade, this wooden panel being designed to catch the mud thrown up by the horses' hooves.

But a consequence of the screen's position was that the glass was now some distance from the driver's seat. The answer came in about 1908 with the arrival of the scuttle (so named because some examples resembled a coal scuttle), which introduced a visual union between the bonnet and the coachwork.

The dashboard was therefore moved back, becoming the bulkhead, which separated the engine from the car's occupants. Initially the instruments moved along with it, but this rendered them virtually unreadable. So they were transferred to a secondary board introduced directly in front of the driver. But what was actually the instrument panel, or fascia, also went by the name 'dashboard'!

The scuttle accommodated the driver's feet and allowed him to sit further forward in the car, so that he or she was now closer to the windscreen. This was now repositioned at the rear of the scuttle. Its arrival marked the last significant change to the design of coachwork to which the French contributed before the outbreak of the First World War in 1914.

This body on a 1914 Peugeot by Rothschild of Paris is notable for its beautifully executed timber-faced finish. Note that the bonnet is in line with the rest of the body and the presence of a scuttle.

SLOWCOACH IN THE FAST LANE, 1896-1914

THE first English Daimlers began leaving the company's converted Coventry cotton mill in 1896 and they effectively marked the birth of the British motor industry. But many of the country's carriage builders did not believe that the automobile represented a threat to their way of life. Because of its noise and lack of refinement, the horseless carriage was looked upon as at best a working vehicle that could serve the tradesman, the commercial traveller or the doctor, who could make more calls than when they were reliant on horse power. The belief was that the carriage would remain supreme for pleasure purposes.

While some coachbuilders could trace their origins back to the sixteenth century, the trade experienced considerable growth towards the end of the eighteenth, because of the prospering economy and the introduction and expansion of the network of turnpike roads. The momentum continued into the nineteenth century, with the aspiring middle classes growing in numbers and influence. The ownership, or even hire, of a carriage was seen as an essential indication of social status.

In 1814 there had been some 23,000 private four-wheel carriages licensed in Britain. This figure doubled to 46,000 in 1834 and soared to 125,000 in 1874, when no fewer than 432,000 carriages were in use. Soon afterwards about 40,000 new horse-drawn vehicles were being built annually, a figure that would not be surpassed by indigenous car production until the early 1920s.

So, as the Victorian era drew to an end, Britain's coachbuilders could reflect on more than two generations of prosperity, the fluctuations of the trade cycle excepted. They were to be found throughout the country, with even the smallest of towns possessing one or more such businesses. It has been estimated that in the final third of the nineteenth century there were some four hundred coachbuilders located throughout Britain, of which about one hundred were based in London. However, other sources put the overall figure at more than a thousand.

Opposite:
The overwhelming majority of coachbuilt bodies produced in the pre-1914 era were built in bodyshops owned by the motor manufacturers. Here tourers are under construction by Sunbeam at its Wolverhampton factory in 1911.

Above left: Open bodies reigned supreme c.1906. Note the limousines in the left background and the rear-entrance tonneau third from left in the foreground with step and back door apparent. The location is the Red Lion Hotel, Henley-on-Thames, Oxfordshire.

Above right: Where is the horse? The Bersey electric taxi cab of 1897 used bodies by Arthur Mulliner or the Gloucester Railway Carriage & Wagon Company, with wheels and suspension also betraying its carriage ancestry.

London was home to the principal firms providing carriages for the aristocracy, who were often resident in town houses in the capital from the eighteenth century onwards. Therefore Barker, established in 1710, Mulliner, London, and Thrupp & Maberly (1760), Hooper (1805) and others were all based there.

However, Rippon, Britain's oldest coachbuilder, had been in business in Huddersfield since 1555 and is credited with the construction of Queen Elizabeth I's coach. Joseph Cockshoot was established in Manchester in 1844, Mulliner of Northampton dated from 1760, and Salmons & Sons had been at its Newport Pagnell, Buckinghamshire, address since about 1830. Windovers of Huntingdon opened for business in 1860, three years before James Young in Bromley, Kent.

It was perhaps inevitable that this most conservative of trades, at a time when many coachbuilders were still making their own paint and washers, would not have perceived the arrival of the motor car as a threat. But the passing, in 1896, of the Locomotives on the Highways Act effectively raised the green flag for Britain to adopt the automobile, ten years after its appearance in Germany.

Many London coachbuilders were members of the Worshipful Company of Coachmakers, which in 1667 had been granted its royal charter by Charles II. In 1898 the Company, together with the Institute of British Carriage Manufacturers, established in 1881, and the London County Council, co-operated with the Polytechnic in London's Regent Street to found the Day School of Carriage Building. It offered an excellent education, both technical and practical, to those about to enter the carriage trade, and the chief draughtsmen of the respected Barker and Hooper companies were to serve as tutors. Such was the timing of the school's foundation that the overwhelming majority of its students were to be employed in the design of motor-car bodies for coachbuilders and as stylists for the industry. In the former category were John Blatchley (Gurney Nutting and Rolls-Royce), James Wignall (Mulliners of Birmingham), Geoffrey Durtnal (Coachcraft) and Alf Ellis, co-founder of Coventry coachbuilders Cross & Ellis. Among stylists to the motor manufacturers were Leslie Hall (Morris), Gerald Palmer (Jowett and BMC) and Ted White (Rootes).

As will have been apparent, the Continentals had set the mechanical and stylistic pace, with Britain being a willing but essentially passive follower. But what British body builders lacked in inventiveness they more than compensated for in the outstanding quality of the finished product, although weight remained a persistent and ongoing concern for the motor manufacturer. As in France, the Roi des Belges line was the dominant style from 1902 until about 1910.

As cars grew in number at the expense of carriages, many hundreds of coachbuilders throughout the country decided to close down rather than meet the challenge of the intruder. Other businesses were formed specifically to cater for the horseless carriage.

Once again London predominated and the most renowned firm was H. J. Mulliner, the motor section of Mulliner, London, founded in 1900, although in 1908 a controlling interest was acquired by John Croall, an Edinburgh coachbuilder established in 1897. In Norwich, Mann Egerton was formed in 1901 while Maltby began building bodies in Sandgate, Kent, in 1902.

However, many of the larger car companies, such as Daimler, Wolseley, Humber, Sunbeam and Rover, would soon establish their own body shops.

Hooper's showrooms were in the heart of Clubland at 54 St James's Street, London, although the works were in Chelsea. The building to the right is the Devonshire Club and many of its members no doubt purchased Hooper-bodied cars.

Even though this 20 hp Thornycroft of 1905 has shaft drive, it echoed French practice in being fitted with rear-entrance tonneau bodywork.

Above left:
The progression from open to closed bodies is well illustrated in these Wolseleys of c.1901. Most Wolseleys were bodied in-house and by 1914 some two hundred men were making forty bodies a week.

Above right:
Stylistically, Daimlers were identifiable by their fluted, nameless radiator, which appeared in 1904 together with the accompanying three-piece bonnet. The curved scuttle dated from 1902, thus anticipating its contemporaries, and pre-dated the arrival of the windscreen.

This permitted them to standardise designs and so reduce costs, and their output dominated the market. In 1910 some six hundred men at Daimler were producing, on average, fifty bodies a week. Chassis could also be supplied to a specialist coachbuilder if the customer desired it but the bespoke nature of the product was reflected by relatively low manufacturing figures.

Hooper, one of Britain's leading coachbuilders of the day, built on average 118 bodies per annum in the years between 1904 and 1914, with output peaking at 168 in 1907. In Manchester Cockshoot body production also attained a record high in the same year, when 118 chassis were so enhanced.

Both companies bodied Rolls-Royces, a business that did not possess a body shop and produced its cars exclusively in chassis form. Rolls-Royce absorbed Bentley in 1931, and these businesses, together with Daimler, would provide the finest coachbuilders of the day with the overwhelming majority of their work.

In 1906 Rolls-Royce introduced Henry Royce's masterpiece, the 40/50 'Silver Ghost', which even its self-effacing creator identified as 'the best thing I have ever done'. The Ghost's chassis was also available with a range of standardised bodies, provided by Barker. Demand came from as far afield as India, where maharajahs soon developed a taste for bespoke English coachwork. This market for more expensive British makes would keep a number of coachbuilders busy well into the economically challenging years of the 1930s.

Of Rolls-Royce's five representatives in Great Britain at this time, two of them, Joseph Cockshoot and John Croall, were coachbuilders. They would invariably recommend their own products to prospective customers, advice that was often, but not always, accepted. Indeed, the prudent coachbuilder would not be wholly reliant on body construction for his livelihood. Those who were destined to survive did so because they took agencies or became distributors for many of the makes of car on the market.

In 1900 the Prince of Wales, later King Edward VII, ordered three Daimlers, and this is one of them, a Hooper-bodied wagonette, able to carry fourteen beaters during the shooting season.

British tastes were in line not only with those of the Continent but also those in the United States of America in espousing what became more generally known as the tourer. In 1912 this open car accounted for no less than 97 per cent of British body production and it maintained its supremacy until the mid 1920s.

Closed bodies, because they contained more materials, were considerably more expensive, and in 1908 Britain made a contribution to their nomenclature that enjoys currency to this day. That year the London Brighton & South Coast Railway began to operate the Southern Belle, later known as the Brighton Belle, hailed as 'the most luxurious train in the world', on account of its opulent Pullman saloons. The word 'saloon' had, since the mid nineteenth century, been accorded to railway carriages devoid of separate compartments. For the payment of a royalty, the Pullman name could be applied to enhance the appeal of a closed coachbuilt body, Rolls-Royce being one example. However, 'saloon' cost nothing and from 1909 onwards it became an increasingly popular description for a closed car for the owner-driver.

Britain had made a more tangible and far-reaching contribution to car styling in 1908. Then the London-domiciled Captain Theo Masui, latterly of the Belgian army and importer of the French Gregoire car, created a flush-sided phaeton and named it the Torpedo. It bore little resemblance to that

The body of Rolls-Royce's legendary 40/50, the Silver Ghost of 1907, is by Barker, a Windham detachable side-entrance tonneau, which permitted the removal of the rear section of bodywork and its replacement with an alternative. To its right is the Silver Shadow, its unitary body construction heralding the end, in 1965, of bespoke coachwork on Rolls-Royces.

Britain's contribution to coachbuilding came in 1908 with Captain Theo Masui's design of his flush-sided Torpedo sporting phaeton for a 14-16 hp Gregoire chassis. It was not the first of its type but it was certainly the most influential.

weapon but was so called because its smooth surface had none of the mouldings so popular at the time. The 1909 Olympia motor show witnessed a rash of similar designs, moving *The Autocar* to comment that it was 'refreshing to find something of British origin'.

The transformation was not completed until the Continentals raised the height of the bonnet to bring it into line with the scuttle and waistline. This enhanced the significance of the radiator, the contours of the body following its profile. (The coachbuilding arm of Masui's business was absorbed in 1913 into the Vanden Plas concern, established in Belgium in 1898.)

Another essentially British contribution was the fixed-head doctor's coupé of about 1910, so named because medical men were among the first regular customers to buy cars and these two-seat closed bodies, with an occasional 'dickey seat' above the rear wheel, were thought to suit their needs particularly well. The saloon apart, it was the only closed body available for the owner-driver.

As will have been apparent, it was possible to buy a Rolls-Royce 'off the peg', as it were, but some members of the aristocracy and those who aspired to their lifestyle delighted in being closely involved in the ordering of their car's body, just as they might commission a suit from Savile Row. Buying a limousine or landaulet could be a lengthy business.

The Hollingdrake Automobile Company of Stockport, Cheshire, began building car bodies in 1904 and was responsible for this rakish two-seater coachwork on a Prince Henry Vauxhall chassis of 1912. It well illustrates that bonnets were becoming in-line with the rest of the body.

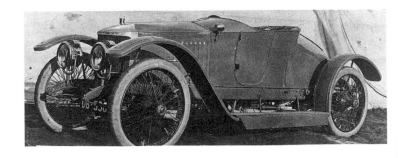

The key player in the transaction was the main dealer, who financed the operation and accordingly carried the risk. He would buy a chassis from the manufacturer and might be responsible for introducing the customer to the coachbuilder and paying the latter's bill.

Drawings would be prepared and the price would be fixed between the manager and his skilled workforce, as in the English furniture-making trade generally. Ideas for styling were shamelessly 'borrowed' from rivals. This was a migratory industry organised around no fewer than thirteen trades, and each quoted for individual aspects of the body's construction. It was an unwieldy process, known as mutuality, but was deeply rooted in the hitherto buoyant carriage trade. The work was also seasonal, usually peaking in the autumn just before the annual motor show.

Sankey's steel artillery wheel displayed at the 1911 motor show. It remained popular on British cars until the late 1920s. In the background is Sankey's pressed-steel body, which seems to have generated little interest.

After approval by the client of a coloured rendering of the finished body, work could begin. The design having been agreed, the body shop would receive working drawings prepared in conjunction with the plans of the chassis supplied by the manufacturer. Plywood patterns would then be made from the drawings and these were sent to the sawmill for the parts to be prepared.

The basis of all coachbuilt bodies at this time was a framework of 'good English ash', although cheaper American imports were already growing in popularity. The frame was reinforced with forged metal gussets and the joints, glued and screwed, were usually lap, half lap, and mortise and tenon. Much of the work was undertaken by hand but, where a uniform body style was being produced, body parts were duplicated on machinery to reduce costs.

The earliest bodies were clad with wooden panels and those that were not curved in more than one plane could be coaxed into the required shape by steaming. The wood was subjected to steam and was then bent around a former or in a jig or curved block. It was then placed, together with its block, in a drying oven. On removal, the wood retained its curvature and was ready for use.

The all-wood frame was usually panelled in mahogany set in white lead to ensure a waterproof joint. However, Roi des Belges coachwork demanded curves, and simple ones required that metal be passed through rollers in much the same way as in 'rounding' a strip to form a wheel rim. The mouldings used to cover external joints might also be machine-made. But when a compound curve was required it was produced by hand, the craftsman hammering out the metal from flat. In 1906 Douglas Leechman noted in *The Car Illustrated* that 'this is rather a new line for the body builder,

This rare photograph, c.1910, shows two landaulet bodies under construction at the premises of Sturt & Goatcher in East Street, Farnham, Surrey. George Sturt was the author of the celebrated book *The Wheelwright's Shop*. In 1920 the business was taken over by Arnold & Comben, another coachbuilder.

and moreover is liable to be disfigured by denting'. Because of this, some of the trickier contours were produced in wood.

When the panels were in place and the body was completed, the process of painting could begin. This was a laborious and time-consuming business, there being '25 to 30 coats ... on a good body'. First came a coat of lead, which took a day to dry, followed by a second and a third. Over the next five days, five coats of 'filling' were applied and each took a further day to dry. Then on went a coat of stain, which required a further two days. Two or three coats of ground colour were then applied, all of which took another two or three days to set. Now the final colour coat was applied and flattened down with powdered pumice, followed by a coat of varnish. Only when two further coats of varnish had been applied was the work considered to be complete.

The above times were minimum ones and Leechman cautioned that a 'coachbuilder of the old fashioned sort is not satisfied unless the time given him for finishing a carriage runs into months'.

In the best traditions of the carriage trade the body was then lined, the worker using a brush with only two or three hairs of great length, a task requiring a steady hand and a good eye. 'Having charged his brush with paint, he lays it on to the work in a way reminding one of a fisherman casting a fly.' The lines were subsequently varnished.

Works coachwork: this commodious Beaufort double cabriolet Daimler dates from 1913. These quiet, sleeve-valve engined cars never wore out and were favoured by the aristocracy in preference to Rolls-Royces, which were considered more suitable for the *nouveaux riches*!

Cunard, established in 1911, was soon purchased by Napier and was responsible for this 'limousine landaulet' on the 40/50 hp chassis. Rolls-Royce's great rival in pre-war days, Napier ceased building cars in 1924.

Next came upholstery, and the seats were stuffed with horsehair and coil springs, the covering usually being buttoned leather for open cars. An alternative was leathercloth, or pegamoid, the trade name for waterproof cloth or imitation leather. Closed cars, not exposed to the rigours of the climate, variably used either corduroy or cloth.

Wings were usually made of metal, but occasionally wood, strengthened with metal, was used, having been steamed into the required shape.

At this time the hood was made of leather or paramatta, a combination of wool and cotton. The hood sticks were shaped by steaming, the metalwork having been produced in the forge, an important facility common to most coachbuilders and a source of many a fire!

From the foregoing it will be recognised that coachbuilding was lengthy, labour-intensive work, and skilled men were at a premium, especially in these days before the First World War. In 1913 a coachbuilder could command a wage of £2 14s (£2.70) for a 47 hour week at a time when 8½d (3.5p) an hour and 54 hours were the norm.

Vincent of Reading, Berkshire, was founded in 1804 and built its first car body in 1899. It was responsible for this sociable berline on a 20 hp six-cylinder Standard chassis of 1911. The lack of front doors did not catch on!

A bodymaker's apprenticeship could last for five, six or more years. His family might be required to make a payment of £10, and another £20 would be required to pay for a tool chest.

In Coventry aspiring coachbuilders underwent a seven-year apprenticeship and, on its completion, from 1910 they were made freemen of the city. By 1917 many of these men would have died in the trenches of the Western Front, but the coming of peace would bring a period of unprecedented growth for the coachbuilder.

HOOPER & Co (COACHBUILDERS) LTD
AGENTS FOR ALL LEADING MOTOR CARS

Motor-Body-Builders and Coachbuilders to

HIS MAJESTY THE KING	⎫ By Royal	H.R.H. THE PRINCESS MARY, Viscountess Lascelles
HER MAJESTY THE QUEEN	⎬ Warrant	H.R.H. THE PRINCESS ROYAL
H.R.H. THE PRINCE OF WALES	⎭ of appoint-ment.	H.R.H. THE PRINCESS VICTORIA
		H.R.H. THE DUKE OF CONNAUGHT
		H.R.H. PRINCE ARTHUR OF CONNAUGHT

54 ST. JAMES'S STREET, PICCADILLY, LONDON, S.W. 1

HEYDAY OF THE
COACHBUILDER, 1919–29

THE insatiable demands for materiel, from aircraft to armaments, needed to produce an Allied victory in the First World War, transformed British industry through the necessity of introducing mass-production methods, which sought to eliminate individual skill from manufacturing processes.

Such techniques were embraced by the more successful British car makers. The accent was on growth and the annual output of the British motor industry trebled from 1923, when some 66,396 vehicles were completed, to 182,347 in 1929. Paradoxically, the overwhelming majority of these cars, from Austin Seven to Rolls-Royce, were fitted with hand-built bodies which were reliant, to a greater or lesser extent, on the manual skills of the craftsmen producing them. But while the little Austin received standardised *bodywork*, the majestic Rolls-Royce Phantom II would have been enhanced by bespoke *coachwork*.

A record sixty-three car makers exhibited their wares at the 1919 motor show, reflecting the wave of optimism that followed the ending of the First World War. Some coachbuilders were forced to offer their employees 5 shillings (25p) per hour to prepare cars in time for the event but such high wages were not destined to last. Twenty years later, in 1939, the industry was not paying more than 2 shillings (10p).

Of Britain's 250 or so significant body makers, no fewer than 131 were established in the years between 1919 and 1925. They included Gurney Nutting and Park Ward (1919) and Freestone & Webb (1923), the last two being based in north-west London, which was to become a centre of the trade. Carbodies and Cross & Ellis (also 1919) were located in Britain's motor city of Coventry, while another Midlands firm was Mulliners of Birmingham, reconstructed in 1924.

Others changed status. Thrupp & Maberly, one of the most respected names in the industry, succumbed in 1925 to a takeover from Rootes Ltd, which was not yet a motor manufacturer but Britain's largest car distributor.

Opposite: By royal appointment, a Hooper advertisement of 1927 illustrating a sedanca de ville body on a Rolls-Royce Phantom I chassis.

Vanden Plas completed this body for an R type 25/85 hp Daimler with Barker wheel discs in December 1926. This advertisement, with St Stephen's Tower at the Houses of Parliament in the background, dates from the following year.

An Enclosed Drive Limousine mounted to a 25/85 h.p. Daimler Chassis

COACHBUILDERS AND MOTOR ENGINEERS

Vanden Plas
(ENGLAND) 1923 LIMITED

KINGSBURY WORKS HENDON LONDON·N·W·9

In view of such activity, it is not surprising that a record sixty-three coachbuilders exhibited their wares in the Carriage Work Section at the 1926 Olympia motor show. Thereafter their numbers began a slow decline.

However, Peter Wharton, Park Ward's post-war stylist, who joined the business in 1934, has pointed out that British coachwork of the inter-war years 'was hard to beat. The combination of oak bottom sides, ash framing, aluminium panels and steel wings, together with

There was always work for coachbuilders during the inter-war years because many of the medium-sized and smaller motor manufacturers produced their cars only in chassis form. This is a 1½ litre Invicta frame built in 1932 and 1933.

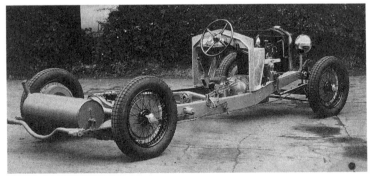

Although suffering from the ravages of time, this photograph, c.1922, shows coachbuilder Enrico (Harry) Bertelli in one of his creations, a saloon body for the Enfield-Allday chassis designed by his brother Augustus (Bert). They would later work together at Aston Martin.

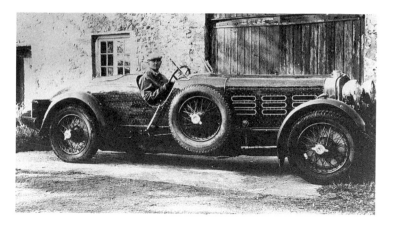

This 1924 Boulogne Hispano-Suiza was originally owned by André Dubonnet of aperitif fame and came to Britain in 1925. The body, made of Brazilian tulip-wood, more often used in furniture construction, with its panels copper-riveted in place, is by French aircraft maker Nieuport-Astra. At the wheel is Rodney Forestier-Walker, who bought the car in 1950.

Connolly Vaumol hides, Ernest Turner's carpets and West of England cloth ensured a lasting elegance' (quoted in *The Rolls-Royce* Wraith by Tom Clarke).

Styles and construction methods initially differed little from pre-war days although aluminium overtook steel in popularity for body panels, with steel beginning to be used for body frames. Duck replaced leather as a hood material. A popular innovation came from Salmons, which introduced the Tickford Winding Hood to its bodies in 1925. By means of an ingenious mechanism, an owner needed only to turn a handle, which was 'as easy... as winding a gramophone'.

Stylistically, the overwhelming majority of cars produced in 1919 would not have looked out of place ten years later. Lacking in external decoration, they were understated, upright and well proportioned.

A small but significant innovation was the arrival of the wind-up window, which began to replace the almost universal pull-up strap. The first of these had appeared in 1913, one example being the Rawlence mechanism, which incorporated a lazy tongs device.

Less obvious to the customer was the introduction in 1922 by Daimler of the body sub-frame. Attached to the chassis by rubber cushions, it isolated coachwork from chassis flexing and was widely adopted throughout the industry.

Much more apparent was the Weymann body and, like so many innovations, it came from France. Charles Terres Weymann, inspired by aircraft construction, in 1921 conceived a flexible, fabric-covered body that by its nature was considerably lighter than its metal-panelled equivalent. It addressed the matter of weight, always a bugbear, and above all was not prone to creaks because the wooden frame was devoid of rigid or closed joints. Instead the timbers were separated by 0.16 inch gaps and secured by metal plates. Indeed, *The Automobile Engineer* observed in 1935 that the Weymann

Weymann body construction, in which wooden members, not touching but separated by metal plates (see right inset), were flexibly connected. The frame was then covered with padding and tightly stretched fabric.

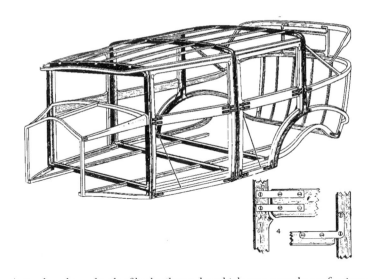

'introduced new levels of body silence, by which even now the perfection of coachwork is measured'.

Based in Addlestone, Surrey, Weymann was responsible for the body of the well-appointed 1928 Hillman Segrave coupé, named after the then holder of the world land speed record. It sold for a pricey £398.

The first of the line was seen at the Paris Salon in 1922 and it crossed the English Channel to appear at the 1923 London event. The coachbuilders paid Weymann a royalty of £12 per car, although Louis Antweiler of Mulliners of Birmingham succeeded in having his figure reduced to £1! But in 1929 the Weymann suddenly went out of favour on the Continent, although it lingered on in Britain for a little longer. It had proved to be difficult to mass-produce and keep clean and it was tricky to repair.

Riley had been a firm adherent of the fabric body and its Nine was immensely influential from a stylistic standpoint. The 1927 Monaco saloon, designed by Stanley Riley, had an inclined windscreen, high waistline in the

Trend-setter: Riley's fabric-bodied Monaco saloon of 1927 by Coventry-based Midland Motor Bodies. Note the high waistline in the Continental manner, inclined windscreen and boot. Bob Porter of Riley agents Boon & Porter of Barnes, London SW13, surveys his substantial cup.

Continental idiom and low roof, this apparent contradiction being made possible by the introduction of foot wells for the rear passengers.

It also featured a boot, then something of a rarity, most cars being fitted with a folding rack at the rear to which a trunk could be attached. This replaced the roof rack of pre-war days. As David Scott-Moncrieff, that celebrated 'purveyor of horseless carriages to the nobility and gentry', has observed: 'By 1930 carrying luggage on the roof of one's car had, for some reason, become as non-U as saying "Pardon" or "Pass the cruet"!'

The Monaco was also significant for being a closed car because from the mid 1920s there was a great change in body styles: the tourer, which had dominated the market since the early years of the century, began to be overhauled by the saloon in popularity. By 1928 the open car was in retreat and that year it accounted for just 38 per cent of British car production. The figure had dropped to 5 per cent in 1930.

The tourer with its hood and side screens could never be regarded as completely weatherproof and the attraction of the saloon was compelling. But more materials were used in its construction and a closed body took longer to build than an open one, and so it was consistently more expensive. The price differential appeared insuperable.

The answer to the problem came from the United States, where the extremes of climate were more pronounced and the case for the weather protection offered by a saloon was overwhelming. It was championed by Roy Chapin of Hudson, whose low-cost Essex line in 1922 introduced the cheap closed car to the United States market. Demand was instantaneous and, in due course, it allowed the price to be reduced.

The success of the Essex was not wasted on Alfred Sloan, president of General Motors, who also recognised that the answer, in part, lay in volume. The other element of the equation was the steel saloon, even though it

The Morris bodyshop at Cowley, Oxford, in 1920, the year of its opening. Construction methods were conventional, with the company producing about eight cars a day. By 1925 the figure had soared to some two hundred and jigs were an established part of the body-building process.

required massive presses and costly dies to mass-produce the panels. But it eliminated the need for the skilled labour required for coachbuilt bodies.

In Britain a complete steel touring body could have been bought in 1911 from Sankey but major developments occurred in the United States, emanating from the Budd Corporation, established in 1912. In 1914 it supplied the new Dodge marque, offering America's first mass-produced car with an all-steel body, and in 1917 the process was extended to the saloon.

In 1919 Hugh Adams, Budd's go-getting salesman, crossed the Atlantic to visit France and England. In consequence André Citroën, France's leading manufacturer, adopted Budd patents and in 1925 introduced the pressed-

Seat squabs are being assembled at Cowley in March 1925. Horsehair, the main constituent of upholstery, is much in evidence, as are the spiral springs in the right foreground.

Painting wings at Cowley in 1920. They were later dipped in an enamelling tank and, ultimately, spray-painted.

steel saloon to the European market. As Britain's largest car maker, William Morris had much to gain from the concept and in 1926, also in association with Budd, he established the Pressed Steel Company at Cowley. It would, in due course, spell the end of the hand-crafted body.

In any event the coachbuilders had problems of their own, which lay in the very nature of their products. Britain had more makes than any other European country, and each used a different chassis, so that any attempt at standardisation was virtually impossible. Furthermore, some manufacturers produced a bewildering variety of cars. In 1927, Daimler, one of the worst offenders, offered twenty-three different models (excluding body styles), five engines and twelve lengths of chassis.

It was against such a background that in 1924 W. Ferrier Brown told the Institution of Automobile Engineers: 'The author cannot call to mind any other trade in which standardisation plays such a small part as in the coach trade. We have standard wood screws and standard sizes for tacks, but beyond that it is absolute chaos.'

Having said that, the trade did begin to address the problem and coachbuilders began to offer 'bespoke' designs that incorporated the same body shell, doors and wings. This was where the volume producers with their in-house body facilities scored by the extensive use of jigs and the paring-down of timber. But even then the build process remained time-consuming and expensive.

A limitation in body manufacturing, be it volume or bespoke, was still the long lead times required to paint a car body. In about 1925, however, cellulose spray paint appeared in Britain, drastically reducing body production times, although the laborious process of brush painting and varnishing continued for a time on the more expensive cars. Like so many

Freestone & Webb was responsible for this drophead coupé body on a Bentley 6½ litre chassis, delivered to E. Bullivant in August 1929. The cycle wings and truncated running boards were in the French idiom.

technical initiatives, cellulose painting originated in the United States. At Oakland, General Motors was the first car company to treat its products in this way, its Fisher-bodied models for 1924 being sprayed with Dupont's Duco cellulose. Initially only one colour was available, which was marketed as 'True Blue', but other shades would soon follow.

Unlike the previous process, which might take months to complete, cellulose, which had to be sprayed rather than brushed, took just six days. Priming was undertaken on the first day, stopping on the second, rubbing smooth on the third, applying the surface-sealing undercoat on the fourth, and three coats of enamel on the fifth. The final day was spent in polishing.

Three bodies being built by Rippon, a coachbuilder of quality and Rolls-Royce retailer for Yorkshire (West Riding). Note the roof in front of the first of them with its distinctive V-shaped peak. A landaulet takes shape behind it.

Formality: a limousine de ville, probably by Barker, on a Rolls-Royce Phantom I chassis. The two-piece windscreen was popular on British cars until the mid 1920s, when it was replaced by a single-piece screen.

One of the most significant new makes to emerge in the early post-war years, certainly as far as the carriage trade was concerned, was Bentley. Destined to become one of Britain's great sporting marques, it triumphed in the Le Mans 24 Hour Race on no fewer than five occasions and is forever identified with the open touring body required by the event. However, with the arrival of its $6^1/2$ litre model in 1925, Bentley was challenging the market dominated by Rolls-Royce and this demanded closed coachwork. Between 1921 and 1931 Bentley produced some 3,024 chassis and the principal supplier was Vanden Plas, reconstructed in 1923, which bodied 676 of them. It was followed by Gurney Nutting (363), H. J. Mulliner (244) and Freestone & Webb (232).

The 1920s proved to be a harsher economic environment than the prosperous pre-war years and in 1922 Rolls-Royce introduced the 20 hp, a smaller model for the owner-driver. A total of 2,940 were built up to 1929, when production ceased, making it the most popular Rolls-Royce of the day.

The company ordered bodies in batches of ten from Barker, its favoured coachbuilder, which allowed it to sell complete cars from its showrooms in Conduit Street, London. Five coachbuilders were responsible for bodying over half of the bodies on the 20 hp chassis, amounting to 1,527 cars. They were headed by Hooper with 449, Barker (440), Park Ward (325), Windovers (166) and H. J. Mulliner (147).

By contrast, Daimler remained a bulwark of tradition, exemplified by its corporate coachwork and that of Hooper, its preferred coachbuilder. Upright and uncompromising, their lofty proportions would have been instantly recognisable before the First World War. They appealed to the ultra-conservative and socially influential 'dowager duchess' market.

For the younger middle-class buyers with an eye to style rather than performance, initially at least, then MG (Morris Garages) served their needs. The marque was the creation of its talented general manager, Cecil Kimber,

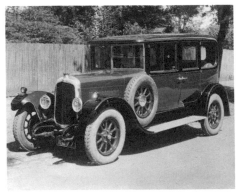

Top left:
Although the body plate says 'The Morris Garages', this MG Saloon de Luxe on a Morris Oxford chassis was built by Carbodies or Raworth. The secondary lever, to the right of the driver, is for the Barker mechanical dipping mechanism and identifies this 1926 car.

Top right:
Anglo-French hybrid: a Morris Garages Saloon de Luxe body, probably by Raworth, on a 1926 Morris-Léon Bollée, powered by a six-cylinder overhead-camshaft Wolseley engine. The sliding front window is an unusual feature.

who combined a marketing flair with an artistic ability to design special coachwork made for him by Carbodies. These bodies enhanced Bullnose and Flatnose Morris chassis and paved the way in 1928 for the popular M-type. Its open two-seater coachwork was fashionable, being fabric-bodied with a pointed tail, for which Carbodies charged MG just £6 10s. (£6.50) apiece. Kimber was also responsible for the design of the one-off bodies marketed by Morris Garages. These were usually built for him locally by Raworth, which had bodied the original Morris Oxford of 1912.

Morris's dominant position in the British market was challenged throughout the 1920s by Herbert Austin, whose most popular product was the diminutive Austin Seven, introduced in 1922. Although it was usually fitted with factory bodywork, nonetheless a variety of coachbuilders offered special bodies on the model, and one of these was the Swallow Sidecar Company, established in 1922 in the seaside resort of Blackpool. Its co-founder was young William Lyons, a stylist of the highest order, who would be counted among the best in the world. His achievement was all the more extraordinary because, like Kimber, he was self-taught, but, unlike him, he could not even draw. Lyons can therefore be regarded as a sculptor rather than an artist, but he was wholly reliant on Cyril Holland, an employee with coachbuilding experience, to interpret his ideas.

In 1927 Lyons introduced the Austin Swallow Seven tourer and followed this in 1928 with a saloon. Recognising that the business needed to be in Coventry, the hub of Britain's motor industry, Lyons moved there in 1928.

Stylistically, sports-car coachwork was unrelated to that of saloons and this was especially so with Vauxhall's 30-98 model. Its works bodies, the Velox and the Wensum, were superior to those of independent coachbuilders.

But Britain's motor industry and the coachbuilding companies that supported it were decimated by the crash of the American stock market in October 1929. Things would never be quite the same again.

Diversity c.1928: C.W. Hayward of Wolverhampton began building car bodies in the early 1920s and obtained contracts from Morris and Rootes. Behind the fabric saloons in the foreground are Guy and Morris commercial-vehicle bodies, and beyond them and on the right in-house AJS saloons and coach bodies.

Ace wheel discs were 'designed to enable wire or artillery wheels to be quickly cleaned...' and smartened up a car's appearance. A rival design was produced by the coachbuilder Barker.

ACE UPER WHEEL DISCS

FOR MORRIS CARS

CORNERCROFT LTD.
ACE WORKS, COVENTRY.
Telephone: 4123 (2 lines) Telegrams 'Discs Coventry.'

Vauxhall's works manager, A. J. 'Jock' Hancock, was responsible for the delectable boat-tailed Wensum body of 1924, named after the River Wensum in Norfolk, where he kept a fast motorboat. Perhaps only fifteen or so 30-98s were so enhanced.

THE HALL MARK

SUNSHINE BEFORE TWILIGHT, 1930–9

THE Depression produced many casualties among Britain's car makers. From a coachbuilding standpoint, the most significant of these was Bentley, which was bought by Rolls-Royce in 1931. But in the longer term the British motor industry was to boom and in 1933 overtook France's to become the largest in Europe. Yet growth was to be coupled with a decrease in the number and influence of British coachbuilders, a decline that accelerated from 1935 onwards. In 1933 a total of forty British companies exhibited at the Olympia motor show; by 1938 the figure had dropped to twenty-six.

Coachbuilding had been given a temporary boost in the period from 1933 to 1937 by the appearance of the so-called Anglo-American hybrids, such as Railton and Brough Superior created by the fitment of English bodies on cheap American mechanicals. But other coachbuilders dropped from automobiles altogether and concentrated on the construction of coach and bus bodies, a market that grew in popularity during the 1930s.

These were the years when Britain's 'Big Six' motor manufacturers emerged and mass-production predominated. In order of market share in 1939, they were Morris, Austin, Standard, Rootes, Ford and Vauxhall, and all embraced the concept of pressed-steel bodywork.

As far as the coachbuilders were concerned, the top-quality bodies still came from Barker and Hooper, despite a tendency towards excessive weight. Gurney Nutting was in this league but it also possessed flair and style, and the same could be said of Vanden Plas. Park Ward's coachwork was rather lighter and some in the trade considered James Young's bodies 'too flimsy'.

Today the coachbuilt body, with some justification, is invariably considered superior to the mass-produced one. But an illuminating perspective, voiced in 1932, highlights the advantages and shortcomings of the hand-built product. 'Coachbuilt bodies which carry on the carriage building tradition... give an impression of solidity and dignity...', a

The exclusive nature of the coachbuilt car as the preferred choice of the rich and well-to-do is the theme of this Dunlop advertisement of the 1930s.

Top:
Gurney Nutting
was responsible
for this sedanca
drophead coupé
on a Rolls-Royce
Continental
Phantom II,
delivered in March
1934 to Madame
Ossarie. Note the
substantial trunk.

Bottom:
La crème de la
crème: Vanden Plas
charged Squire
£195 for this body
with a metal frame
rather than the
usual wooden one.
Styled by Adrian
Squire, it was
delivered in
February 1935 and
was to serve as
the company's
works
demonstrator.

commentator observed. The downside was that they 'develop rattles in time, however well they are constructed...'.

As will have been apparent, design in the 1920s had been remarkably static. But if a 1930 car were parked alongside its equivalent from 1939, the difference between the two would be striking. The earlier body was upright and angular; the later one lower, wider, curvilinear and more elongated. By then the closed body was supreme; indeed, in 1938 no fewer than 98 per cent of new cars sold in Britain were saloons.

So the perpendicular look of the 1920s was replaced by sloping profiles suggestive of movement. One of the first casualties was the deletion of the projecting peak above the windscreen, to be replaced by a curved, more 'eddy-free' front.

In the intervening years the science of aerodynamics had moved to the fore although in Britain it was more discussed than applied. This was because, historically, tradition overrode innovation and in any event a body with wind-cheating properties that was truly efficient aerodynamically was visually unattractive and therefore unlikely to find buyers. This did not prevent British

car companies from introducing elements of the new approach, just one feature being the extended tail, which conveniently incorporated the increasingly popular boot. A rash of names, such as Aero, Aerodynamic, Aerodyne, Aeroflow and Airline, were suggestive of the new genre.

As far as the coachbuilders were concerned, 'the gentlemen of impeccable taste', as Park Ward's Peter Wharton described them, who ordered bespoke coachwork were invariably conservative in outlook and liked the reassurance of the established order. They were being served by a similarly conservative industry, which was in decline and unlikely to take risks.

Yet quality from the best companies was second to none. The foreman at Gurney Nutting memorably told a customer, complimenting him on the excellence of the firm's interior woodwork, that he could have 'made him a veneered waistcoat if he had wanted one'.

Continental coachbuilders, such as Binder, Kellner and Labourdette, were on a par with their British counterparts. But other European houses were hampered by the lack of domestically grown ash timber and tended to use beech, which lacked its elasticity and bending properties.

Top:
This coupé de ville body Is just one of many magnificent illustrations to feature in a catalogue by French coachbuilder Gallé, which bodied Hispano-Suiza and Stutz. The publication was owned by Percy Twigg, who established Coachcraft in west London in 1934 and, like many British coachbuilders, was much influenced by the French.

Bottom:
That aerodynamic look: Vanden Plas built its pillarless silent-travel saloon principally on Alvis and Bentley chassis. This 3 1/2 litre example of the former, in blue and black with cream wheels, was delivered in March 1936.

That is not to say that Continental styles were not admired in Britain. Although David Scott-Moncrieff has written that the Parisian coachbuilder Figoni & Falaschi was dismissed for being 'phoney and flashy', Wharton has recalled that 'quite a number of people thought Figoni and Chapron bodies on the D8 Delage were the finest expression of the coachbuilder's art'.

So change was in the air and streamlining was the catchword of the 1930s, not only being applied to cars but to railway locomotives, household appliances and buildings. However, few examples of so-called 'streamlined' British coachwork had been subjected to wind-tunnel testing. Their creators were reliant on the established parameters of instinct, eye and manual dexterity, what American stylist John Tjaarda called 'guessamathics'.

As in the past, these influences emanated from overseas, but this time they came from the New World as well as the Old. The United States offered the unsuccessful wind-tunnel-honed Chrysler Airflow of 1934 and the more conventional but still radical (and more popular) 1936 Lincoln Zephyr.

Germany was the true home of aerodynamics, following pioneering work in 1921–2 by former aero engineers Paul Jaray and Edmund Rumpler, who were forced to divert their talents from aircraft by the Treaty of Versailles. German Grand Prix racing cars from 1934 onwards would benefit, and road cars similarly reflected such influences. The Continentals also embraced the concept of the sports coupé, as it was aerodynamically more efficient than an open car.

All this activity was principally due to the fact that Europe's motor industry tended to recruit graduates direct from technical universities, where they had been exposed to the frontiers of technology. Britain, by contrast, was mainly reliant on apprenticeships, which consolidated the status quo.

In 1935 Maltby was offering this Redfern 'Saloon-Tourer', with the refinement of a hydraulically actuated hood. It was available on the pillarless four-door and two-door 'coupé-tourer'.

Brooklands Flying Club provides a backdrop to a 1936 Railton Sportman's Coupé by Ranalah, active in south-west London in 1935–9. The flap adjoining the semaphore indicator gave access to a compartment to accommodate a set of golf clubs, an optional provision.

Such initiatives would be accelerated by the arrival, in 1933, of the German autobahns, and France, with its string-straight roads, was also receptive to aerodynamic influences. The results of both countries' experiments were displayed at the Paris Salons of the 1930s and, as in previous years, these shows would also provide British coachbuilders and industry stylists with many ideas, some good, but others unashamedly outlandish.

However, in 1935 came a British initiative, no less, when Freestone & Webb offered its brougham on a $3^1/2$ litre Bentley chassis in what was then known as the 'top hat' style and is today universally known as the 'razor edge'. Although it was widely copied and endured into the post-war years, there was, inevitably, a French precedent. Back in 1922, Labourdette had offered its angular, crisply executed Art Deco saloon on a Silver Ghost chassis.

At the 1938 motor show, the last to be held before the outbreak of the Second World War, *The Autocar's* comments gave no inkling of an industry in its death throes. 'Certainly only one conclusion is inevitable', it opined: 'that the products of British specialist coachbuilders are unsurpassed...'

Some change had been taking place, however. In 1933 Rolls-Royce turned to Park Ward as the preferred supplier of coachwork for its newly acquired Bentley marque. It therefore took a financial interest in the business and went on to order saloon bodies in batches of twenty and thirty at a time.

Yet behind the scenes Rolls-Royce was soon expressing concern with Park Ward's composite coachwork, which was regarded as being of poor quality and durability. Matters had come to a head in 1934 when Rolls-Royce's works manager Arthur Wormald took delivery of a new Park Ward-bodied

One of the first all-steel Park Ward-bodied Bentleys, this 4 1/4 litre car was delivered in November 1936 to Thomas Ward, of the Sheffield steel family. It is owned by Will Morrison, who was responsible for its restoration.

20/25 saloon, only to find that its doors flew open when it was on the move. There were similar instances of coachwork problems with customers' cars.

As a consequence, in 1936 Park Ward announced its all-steel body, which obviated the need for the traditional ash frame. Tooling up for the steel framework could be justified by the Bentley contract although, as will have been apparent, the feature was far from new.

At least one coachbuilder/manufacturer would follow Park Ward's example in concept, if not detail. William Lyons had in 1931 launched the SS marque and his Jaguar saloons for 1938 featured all-steel bodies, which eliminated the need for the time-consuming and therefore expensive work in the corporate sawmill. Small pressed-steel panels were produced in the company's Coventry works; Rubery Owen contributed doors, while quarter panels came from Sankey, Pressed Steel and others.

Park Ward's steel frame of 1936, the tooling of which was justified by Rolls-Royce's allocation of the overwhelming majority of its Bentley chassis to that coachbuilder. A total of 1,066 bodies were built, some 40 per cent of the 3 1/2 and 4 1/4 litre chassis.

With the demise of coachbuilding skills, the work of the lead loader increased. He would heat the metal with a blowlamp and then spread it using a wooden paddle to conceal spot welds and tricky bends. Within six years Lyons had made the transition from traditional coachbuilder to modern body builder. Such investment was made possible by rising sales of the SS Jaguars.

This process was not applied to SS's low-production 100 open two-seater. And, as far as British sports cars were concerned, Italy, which in the 1920s had been content to follow convention, emerged as a stylistic force in the following decade. Alfa Romeo's open two-seaters were enhanced by exquisite designs from Zagato and by Touring of Milan, perhaps the greatest of them all. Such Italian themes soon found expression in such British sports cars as the Riley Imp, Sprite and MPH, MG and the rare, delectable Squire. Its lines provided an inspiration for the Morgan 4/4 of 1935, a newcomer to the ranks of Britain's performance cars from a manufacturer who had begun in 1910 by building motorcycle-engined three-wheelers.

Morgan represented a challenge to MG, which continued to dominate the market, partly because Kimber maintained his stylistic flair, aided by catalogue illustrator Harold Connolly. His J2 of 1932 with its double-humped scuttle, cutaway doors and flowing wings provided a distinctive and influential line that was to endure until 1955.

Kimber benefited from being safely within the Morris Motors empire and it might be imagined that, with the growth of the big battalions, there was little room for the independent coachbuilder. But some of them flourished by becoming allied to a mass producer. As that perceptive motoring historian Michael Sedgwick has pointed out, 'the ladies were driving themselves to tennis or bridge and they were not enchanted by the duller body styles imposed by the spread of presswork'.

HUMBER "PULLMAN" LIMOUSINE

The chauffeur-driven Humber Pullman Limousine of 1932 had a coachbuilt corporate body at £735, but a sedanca de ville version by in-house Thrupp & Maberly would cost £1,095.

Cunard became Morris's favoured coachbuilder of the 1930s, having been purchased in 1931 by that company's London distributor, Stewart & Ardern. It was responsible for this Ten Special Coupé of 1934.

John Blatchley, who became Gurney Nutting's chief draughtsman at the age of twenty-three in 1936, was one of Britain's outstanding stylists. His sedanca de ville (design 189) for Princess Mdivani on a Rolls-Royce 25/30 chassis dates from 1936.

In consequence the car makers found places in their catalogues for coachbuilt variants of their mainstream models. Therefore Morris used Cunard, Austin went to Gordon, Standard to Avon, and Vauxhall to Grosvenor. Rootes already owned Thrupp & Maberly, and its big Humbers were duly enhanced, although Carbodies served the needs of the Hillman Minx.

In 1932 Wolseley had introduced its Hornet sports model and no fewer than 129 variants by ten coachbuilders were offered in 1931 and 1932 alone. Jensen, established in 1928 at West Bromwich, produced a smart drophead coupé.

With the Hornet's departure in 1935, Vauxhall became the recipient of a range of cheaper coachbuilt bodies.

Such diversification was possible because all of these manufacturers retained a separate chassis. But this was to be dispensed with on the Vauxhall 10 for 1938, which was the first British model to feature unitary construction. It foreshadowed a universal change that would add to the coachbuilder's woes.

Rolls-Royce's Phantom III of 1935 was not favoured by coachbuilders because its radiator was mounted further forward than those of its predecessors, so unsettling proportions. However, H. J. Mulliner has done its best with this Jack Barclay-designed razor-edged sedanca de ville of 1937.

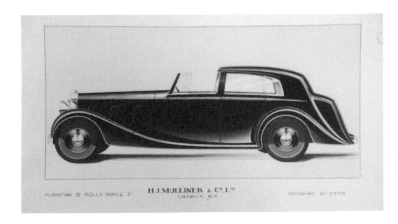

PHANTOM III ROLLS-ROYCE F   H.J MULLINER & Cº Lᵗᵈ   DRAWING Nº 5974
                             CHISWICK W 4

Even the best names were struggling for survival, either closing or being absorbed by larger organisations. Their plight was exemplified by Park Ward. Despite its commitment to Rolls-Royce, it recorded a loss of £2,658 for the first six months of 1938 and in the following year was purchased by the makers of 'the best car in the world' for a total investment of £82,000.

BSA, custodian of Daimler, followed its example in 1940 by acquiring Hooper – which had already in 1938 bought the old-established Barker company from the receiver.

This left H. J. Mulliner somewhat isolated in London as an independent, although it did have Freestone & Webb and Vanden Plas for company. In 1937 the prestigious retailer Jack Barclay personally acquired James Young and in 1945 purchased Gurney Nutting; this was effectively the end of that great name. Some coachbuilders, such as Cross & Ellis, had closed down in 1938, followed by the likes of Coachcraft, Cockshoot and Ranalah. Arthur Mulliner of Northampton succumbed to the motor trade, being acquired in 1940 by Henlys, and in 1944 Raworth was bought by Morris Garages.

There were some survivors. Mulliners of Birmingham, with some 1,500 employees at its height, was well placed geographically to enjoy a decade of relative buoyancy, due in part to the closure, in 1929, of Daimler's bodyshop.

E. D. Abbott of Wrecclesham, Farnham, Surrey, produced 80 per cent of the bodies for Lagonda's Rapier of 1934–5 and, as this advertisement indicates, also sold the car itself.

Freestone & Webb
was responsible
for the coachwork
of this touring
saloon on a
Daimler 4 litre
straight-eight
chassis of 1939.

Mulliners became Daimler's in-house coachbuilder, also serving Lanchester, which BSA had bought in 1931. But Mulliners' profit in 1939 amounted to only £15,030. Other Midlands survivors included Carbodies (Singer and Standard) and Charlesworth (Alvis and MG).

Salmons had found there was a steady demand for its Tickford drophead coupés, particularly from MG, Vauxhall and Rover. Between 1936 and 1939 some two thousand bodies were built. In Farnham, Surrey, Abbott, a newcomer, which had taken over the bankrupted Page & Hunt business in 1929, proved to be another survivor. It was consolidated by a contract from Lagonda for its Rapier, followed by commissions from Talbot, Aston Martin and Atalanta.

But at best body production could be counted only in hundreds. Hooper output reached an all-time high of 316 in 1934 but it dropped to 121 in 1938, and the respective totals for H. J. Mulliner were 126 and 90. Thrupp & Maberly built 211 bodies in 1935, a figure that fell to 144 in 1938. Freestone

Most Lagondas
were bodied in-
house and from
1935 had an
outstanding stylist
in young Frank
Feeley. This
drophead coupé
perfectly
complements the
V12 chassis
designed by W. O.
Bentley and
presented Rolls-
Royce with an
unwelcome
challenge.

& Webb produced 68 bodies in the former year and only 52 in the latter. The 74 bodies that Gurney Nutting built in 1936 echoed the highs of 1925 and 1928 but thereafter totals fell dramatically.

At the heart of this decline was the fact that fewer people were buying Rolls-Royces, Bentleys and Daimlers, which in the late 1930s all experienced falls in production. A revitalised Lagonda, its cars enhanced with impressive in-house coachwork designed by the estimable Frank Feeley, also provided a significant challenge. And American imports such as Packard and Buick offered comfort, speed and reliability at relatively low cost.

The growth of value-for-money sports saloons, first from SS and then from MG, challenged Bentley by offering visually impressive, well-appointed bodies that had previously been the preserve of the coachbuilder. It is no accident that William Lyons and Cecil Kimber both had an intimate knowledge of the business. For those individuals who thought SS was too flashy, an alternative was provided by Rover, which became a byword for understated quality.

As the decade drew to a close, there could be little doubt that the coachbuilding concept as it then existed was not viable. In addition to changes in buying habits, there were the low volumes, the labour-intensive nature of body construction, the archaic method of pricing and, crucially, the bespoke nature of the product.

Coachbuilding, a business that had rarely innovated, could not sustain itself, and this elegant, beautifully appointed but already unstable house of cards all but collapsed on 3 September 1939, when the Second World War began.

This SS Jaguar saloon of 1937 was styled by its chairman, William Lyons, and built with pressed steel panels on an ash frame. In the following year SS progressively switched to all-metal construction.

THE END OF THE ROAD, 1945 TO DATE

Vauxhall's switch to unitary construction in 1938 was soon followed by the other 'Big Six' motor manufacturers. Morris followed in 1939 and Hillman in the year after that. After the Second World War, Ford went unitary in 1950, Austin in 1951 and Standard in 1953. Coachbuilt variants on mainstream models thus became a thing of the past.

Yet another challenge to traditional coachbuilding came from the wartime aviation industry. The so-called 'rubber press', an American import, in which a rubber mattress was used in conjunction with cheap Kirksite tooling, lent itself to the production of low-volume body parts. The Triumph (as SS became in 1945) Roadster of 1946 was one of the first models to benefit from this process and it was subsequently adopted by the likes of Aston Martin, Austin Healey, Jaguar, and Rolls-Royce.

The aircraft industry also provided light alloy extrusions, which continued to replace the timber frame in the more expensive bodies. Hooper adopted such frames from 1949; H. J. Mulliner followed suit in 1952, and James Young the year after.

In the short term at least, the pent-up demand for cars caused by the war resulted in the ranks of the coachbuilders being swollen by some sixty new recruits. There was work for small concerns, and such new makes as Healey, Bristol and Lotus and a revived Allard kept the independent body builders busy for a time. Their ranks were also given a brief and unexpected fillip in 1946–7 when a loophole in the Purchase Tax laws produced a rash of timber-framed estate cars. These persisted, particularly on Alvis and Lea-Francis chassis, even after this tax exemption was closed.

By 1960 the overwhelming majority of these names had disappeared but a notable exception was Harold Radford, who in 1948 first produced an estate body on the Mark VI Bentley. But most of the established businesses had gone. Freestone & Webb closed in 1958, although James Young survived until

Successor to Carbodies, London Taxis International is one of the few survivors of Britain's coachbuilding industry. Here a London taxi, which retains the traditional chassis, is taking shape at LTI's Coventry factory.

Top:
Dead ends: Invicta was briefly revived after the Second World War and the result was the complex Black Prince of 1946. This drophead coupé, photographed en route to South Africa, was by the little-known and equally short-lived London-based coachbuilder Ronald Kent.

Upper middle:
Norfolk-based Duncan Industries offered the same coachwork, styled by Frank Hamblin, on the Healey and Alvis TA14 chassis. Active in 1947–8, its work was centred at the wartime Swannington aerodrome.

Lower middle:
The brief post-war popularity of the timber-framed estate car even extended to Rolls-Royce. Freestone & Webb was responsible for this example on the Silver Wraith chassis delivered to F. C. Tetley in 1950.

Bottom:
Post-war Gurney Nutting bodies were produced by James Young (both companies being owned by Jack Barclay) and styled by A. F. McNeil, whom John Blatchley succeeded at Gurney Nutting. This sedanca coupé graces a Bentley Mark VI chassis.

A contemporary colour photograph of the Bentley Continental, introduced in 1952. Rolls-Royce's Ivan Evernden was the stylist and H. J. Mulliner was responsible for the metal-framed body.

1967. H. J. Mulliner continued but under the ownership of Rolls-Royce.

Paradoxically, Rolls-Royce had played a role in the demise of the coachbuilders by its decision in 1944 to join the rest of the motor industry and switch to pressed-steel bodywork. As far as W. A. Robotham, Rolls-Royce's deputy chief engineer, was concerned, 'it was clear that not only would coachbuilt bodies made in the pre-war manner be of doubtful quality ... they would also be prohibitively expensive'.

To reap the cost benefits, five thousand bodies of uniform design would need to be ordered, a quantity far beyond the scope of any coachbuilder. However, the pressed-steel body was fitted not to a Rolls-Royce, which continued with bespoke coachwork, but to the Bentley Mark VI of 1946, and it was sold as the company's first complete motor car.

The chassis was also made available for specialist coachbuilders and in 1952 came the Bentley Continental coupé. Styled by the company's Ivan Evernden and built by H. J. Mulliner, using its so-called 'stressed skin' construction, it survived in S1 form until 1959. The same year, Mulliner, at its own request, was taken over by Rolls-Royce, and in 1961 it was merged with Park Ward to create Mulliner, Park Ward. Finally, in 1991, it left north London and was transferred to Rolls-Royce's Crewe factory. Today the Mulliner name lives on in enhanced versions of the Bentley Arnage, some 90 per cent being bespoke to varying degrees.

In the meantime, in 1956, Rolls-Royce and Bentley had become allied with the arrival of

Craftsmanship at its finest, Rolls-Royce's 'Flying Lady' motif featured on the interior woodwork of the Crewe-built 'Last of Line' series of Silver Seraphs, of which 170 examples were completed in 2002.

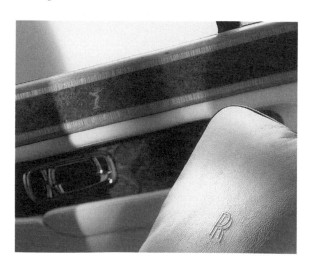

COACHBUILDING

their respective Silver Cloud and S1, which shared the same pressed-steel saloon body. Coachbuilding traditions were maintained because the stylist was John Blatchley, formerly of Gurney Nutting, who had joined Rolls-Royce.

The next generation of cars, the Silver Shadow and Bentley T1 of 1965, was also the work of Blatchley. After sixty years, the separate chassis was banished from Rolls-Royce products, the exception being the low-volume Phantom VI, until it too ceased production in 1992.

H. J. Mulliner was by no means the first major coachbuilding company to be taken over by a car maker. For in 1946 Austin had absorbed Vanden Plas, initially to produce its low-volume Princess saloon. In due course it was drawn into the morass that was British Leyland and its north-west London premises closed in 1979.

By then Ford was dominating the mass market. But back in 1953 it had acquired Briggs Motor Bodies, its supplier since 1932, an initiative that provided the first link in a chain of events that was to change the status of Britain's body makers and some of its surviving coachbuilders.

As a consequence of Ford's action, the British Motor Corporation, a merging in 1952 of the Austin and Morris interests, in 1953 acquired pressed-steel body maker Fisher & Ludlow, established in 1920. Its customers included Standard, which, faced with having a body supplier owned by a competitor, in 1954 secured an agreement with Mulliners of Birmingham for its entire body output. This deal was consolidated in 1958 when Standard bought the business.

Mulliners' 1954 commitment threatened Alvis and Aston Martin, which were both dependent on it for their coachwork. Alvis was eventually to find a saviour in Mulliner Park Ward, which supplied its bodies until Alvis ceased car production in 1967.

Aston Martin's owner, David Brown, responded in 1955 by acquiring Tickford (as Salmons was now known), which was already building some of its bodies, and his company moved into its factory in Newport Pagnell, Buckinghamshire. It was to be Aston Martin's home until 2007.

Morris's Coventry-based Bodies Branch, with origins rooted in coachbuilders Hollick & Pratt established in 1913, was responsible for producing the body of Riley's 1½ litre of 1945. It retained a timber frame, clad with pressed-steel panels, and featured this distinctive fabric roof.

50

Triumph's 1800, renamed Renown in 1949, revived the razor-edged style of the mid 1930s. It was built by Mulliners of Birmingham, but who the stylist was is less certain. It was probably James Wignall, the firm's chief stylist, but Leslie Moore, the chief draughtsman is another contender. This is the limousine version for 1952.

Brown had bought Aston Martin from Gordon Sutherland, who in his turn purchased Abbott in 1950. It proved to be an unlikely survivor, having in 1953 secured a contract with Ford to produce the Farnham Estate body on the Zephyr and Consul models. This activity continued until 1972, when all Ford's estate cars were brought in-house, and Abbott closed that year.

Convertible versions of Ford's Consul and Zephyr had been trimmed in Coventry from 1953 by Carbodies, a line that survived until 1963. But the change in Mulliners' status also affected Daimler and in 1954 Carbodies was bought by its BSA parent. Its mainstream activity was the production of London taxi cabs, and it still functions in Coventry as one of the last survivors of Britain's coachbuilding industry, producing some 2,700 taxis a year.

Daimler, that other champion of the coachbuilding tradition, in 1946 introduced the DE36, Britain's last production straight eight. Invariably Hooper-bodied, it served as the basis of most of the 'Docker Daimlers', which so enlivened the post-war Motor Shows. The company, a case study in corporate inefficiency, also produced a bewildering number of low-volume six-cylinder models and in 1960 was bought by Jaguar. Hooper had closed in 1959.

For a time in the 1950s it looked as though a new material, moulded glass fibre, had usurped the wood/metal combination for small production runs. But its finish lost its sheen with the passage of time, the material proved difficult to repair, and its success was short-lived.

Former coachbuilder Jensen switched to glass fibre in 1953 with its 541 grand tourer although the body contract for the Austin Healey, BMC's corporate sports car, paid the bills. When this was discontinued in 1967, it could only be a matter of time before Jensen closed down and this occurred in 1976.

Stylistically, sports cars continued to reflect Italian themes, which were underpinned when Frazer Nash 'liberated' one of the Touring-bodied BMW

Top left:
Lady Docker, wife of Sir Bernard, chairman of BSA, which owned Daimler, was responsible, with Osmond Rivers of Hooper, for the Docker Daimler show cars of 1951–5. Her dress by Madame Fanny Chiberta featured 'Stardust', the first of the line.

Top right:
The coachbuilder who became a motor manufacturer, Jaguar's Sir William Lyons, arguably Britain's finest stylist, pauses in his work creating a body at Jaguar's factory at Browns Lane, Allesley, Coventry, in the 1960s. In the background is his chief engineer, William Heynes.

328s that had run in the 1940 Mille Miglia. Not only did it directly influence the Jaguar XK120, but also the MGA and Triumph TR2.

A symbolic end to coachbuilding came in 1974 when the National Union of Vehicle Builders, whose origins reached back to 1834, and the United Kingdom Society of Coachmakers merged with the Transport and General Workers Union. Happily, coachbuilding skills remain alive in the old-car restoration business, which still produces ash-framed bespoke bodies to a standard worthy of some of the finest coachbuilders of the past. Funeral directors require hearses and limousines and Coleman-Milne, one of the leading practitioners of the craft, was founded as recently as 1953, having absorbed the old firm of Woodall Nicholson.

As far as the motor industry is concerned, Morgan, one of Britain's oldest surviving car companies, not only stylistically reflects the 1930s, but it continues to build its cars in a traditional way. Morgan still uses English ash, a material that harks back to the days of horse-drawn carriages, which is where our story began.

Today Morgan continues to build its cars with ash framing in the traditional manner. This is a current view of its factory in Malvern Link, Worcestershire, but it could have been taken before the Second Word War.

GLOSSARY OF COACHBUILDING TERMS

These require some qualification. Not only did their meanings change over the years but, confusingly, coachbuilders tended to apply them freely to their styles simply to sound innovative!

All-weather: a body having metal-framed glass winding windows and which offered the protection of a saloon but retained a folding hood.

Berline: a European name for a saloon, derived from a horse-drawn coach that could accommodate four people.

Cabriolet: a term that replaced *all weather* in the 1920s. It is in use on both sides of the English Channel, the German version being rendered as *Kabriolett*.

Cabriolet de ville: refers to when a hood was partially unrolled so that the driving compartment was open to the elements but the rear section remained closed.

Close-coupled: applied to a saloon in which the back seats were very close to the front ones and thus came within the car's wheelbase.

Convertible: a term now universally applied to the *all-weather* body.

Coupé: a short or close-coupled body with windows positioned above its two doors.

Coupé de ville: similar to a *cabriolet de ville* but with the rear portion of the roof fixed rather than movable. The two-door version might be named a *sedanca coupé*.

Dickey seat: a secondary, if draughty, 'mother-in-law's seat' located behind and separate from the driver and passenger. Popular in the 1920s, it went out of fashion in the 1930s.

Doctor's coupé: a two-door, two-seater body with no side windows and a fixed or folding roof. It appeared in Britain in the years immediately prior to the outbreak of the First World War. Popular and so named because of being favoured by doctors, the accent was on ease of access and manoeuvrability.

Drophead coupé: 'head' in this context meant 'roof', so the term refers to a *coupé* with a folding roof.

Enclosed drive: a name usually applied in the pre-1914 era to a body in which the driver enjoyed the provision of a roof, doors and side windows.

Fixed-head coupé: a *coupé* with a fixed roof.

Landaulet: popular before the First World, a *limousine* in which the rear section of roof could be lowered.

Light: the windows of a car were usually referred to as lights, so a six-light saloon is one with three windows on either side.

Limousine: a roomy *saloon* with a secondary fixed bulkhead behind the front seats that incorporated a glass partition to separate the chauffeur from the occupants.

Limousine de ville: a *limousine* in which the roof of the driving compartment could be folded or slid backwards.

Phaeton: originally a four- or five-seater *tourer* of pre-First World War days, and for a time synonymous with *tonneau*. It was popular in the United States between the wars and survives there.

Roadster: an American term that since the Second World War has been used to describe an open two-seater sports car.

Saloon: originally used to describe an owner-driver *limousine* that accordingly lacked a division. Now universally applied to a closed body with a fixed roof, two or four doors and four or six *lights*.

Sedanca de ville: an elegant and impressive body, usually with four doors and *lights*, a partition and a driving compartment with a roof that could be opened if desired.

Tonneau: a four-seater body dating from the earliest days of motoring, in which passengers gained access to the back seats via a door in the back panel. Subsequently applied to the side-entrance version. Today it survives in the term 'tonneau cover'.

Torpedo: designed in 1908, it referred to a body with a sporting flavour with smooth contours, and bereft of decoration.

Tourer: an open body with either two or four seats, and with folding hood and windows.

Touring limousine: a *limousine* with a substantial boot, making it suitable for continental touring.

Established in 1919, Duple of North London switched to coach bodies but continued to body some cars in the 1930s, by then based in Hendon. This touring body is appropriate for a Canadian-built right-hand-drive Buick chassis of 1933.

FURTHER READING

Bardsley, Gillian. *Vintage Style: The Story of Cross and Ellis, Coachbuilders*. Brewin Books, 1993.

Barraclough, R. I., and Jennings, P. L. *Oxford to Abingdon*. Myrtle Publishing, 1998.

Beattie, Ian. *The Complete Book of Automobile Body Design*. Haynes, 1977.

Bird, Anthony. *The Motor Car 1765–1914*. Batsford, 1960.

Clarke, Tom C. *The Rolls-Royce Wraith*. John Fasal, 1986.

Fasal, John. *The Rolls-Royce 20*. John Fasal, 1979.

Georgano, Nick (editor). *The Beaulieu Encyclopedia of Coachbuilding*. Stationery Office Books, 2001.

McLellan, John. *Bodies Beautiful*. David & Charles, 1975.

Munro, Bill. *Carbodies*. Crowood, 1998.

Mynard, Dennis C. *Salmons & Sons*. Phillimore, 2007.

Nockolds, Harold (editor). *The Coachmakers*. Ian Allan, 1977.

Portway, Nicholas, *Vauxhall 30-98: The Finest of Sporting Cars*. Wensum Publishing, 1995.

Robotham, William. *Silver Ghosts and Silver Dawn*. Constable, 1970.

Scott-Moncrief, David. *The Thoroughbred Motor Car*. Batsford, 1963.

Sedgwick, Michael. *Cars of the 1930s*. Batsford, 1970.

Smith, Brian. *Vanden Plas*. Dalton Watson, 1979.

Walker, Nick. *A–Z of British Coachbuilding*. Bay View Books, 1997 and 2007.

Most British motor museums contain examples of coachbuilt cars. Indeed, this will include any pre-1914 car and many of those produced by small to medium manufacturers of the inter-war years.

If you have an interest in coachbuilding, or any other aspects of motoring history, then the Society of Automotive Historians in Britain would welcome you. The membership secretary is Mark Morris.
e-mail: pursang51@yahoo.com

INDEX

Page numbers in italic refer to illustrations

Printed and bound by CPI Group (UK) Ltd, Croydon, CR0 4YY

11/10/2024

01043560-0004